Frank Yeigh

The Rainy River District, Province of Ontario, Canada

A description of its soil, climate, products, agricultural capabilities and

timber and mineral resources, together with the laws pertaining to free

grants and homesteads

Frank Yeigh

The Rainy River District, Province of Ontario, Canada
*A description of its soil, climate, products, agricultural capabilities and timber and
mineral resources, together with the laws pertaining to free grants and homesteads*

ISBN/EAN: 9783744785457

Printed in Europe, USA, Canada, Australia, Japan

Cover: Foto ©berggeist007 / pixelio.de

More available books at **www.hansebooks.com**

RAINY RIVER DISTRIC

PROVINCE OF ONTARIO, CANADA.

A DESCRIPTION OF ITS SOIL, CLIMATE, PRODUCTS, AGRICULTUI
CAPABILITIES AND TIMBER AND MINERAL RESOURCES,

TOGETHER WITH

HE LAWS PERTAINING TO FREE GRANTS AND HOMESTEADS; TO MI
AND TO THE PRESERVATION OF FORESTS FROM
DESTRUCTION BY FIRE.

COMPILED BY

FRANK YEIGH,

OF THE DEPARTMENT OF CROWN LANDS.

BY DIRECTION OF

HON. A. S. HARDY,

COMMISSIONER OF CROWN LANDS FOR THE PROVINCE OF ONTARIO,

1890.

TORONTO:
PRINTED BY WARWICK & SONS, 68 AND 70 FRONT STREET WES
1890.

THE

RAINY RIVER DISTRICT,

PROVINCE OF ONTARIO, CANADA.

A DESCRIPTION OF ITS SOIL, CLIMATE, PRODUCTS, AGRICULTURAL
CAPABILITIES AND TIMBER AND MINERAL RESOURCES,

TOGETHER WITH

THE LAWS PERTAINING TO FREE GRANTS AND HOMESTEADS; TO MINING,
AND TO THE PRESERVATION OF FORESTS FROM
DESTRUCTION BY FIRE.

COMPILED BY

FRANK YEIGH,

OF THE DEPARTMENT OF CROWN LANDS.

BY DIRECTION OF

HON. A. S. HARDY,

COMMISSIONER OF CROWN LANDS FOR THE PROVINCE OF ONTARIO,

1890.

TORONTO:
PRINTED BY WARWICK & SONS, 68 AND 70 FRONT STREET WEST,
1890.

CONTENTS.

THE RAINY RIVER DISTRICT.

Introduction.

The latest addition to the territory of Ontario is the Rainy River District awarded to the Province by the decision of the Privy Council dealing with what is known as "the disputed territory case." With an area of 100,000 square miles, possessing vast tracts of arable soil and merchantable timber, with valuable mineral deposits, with great rivers and lakes, almost all of which are navigable, it is an addition to Ontario's resources the importance of which few comprehend and of which every Canadian may well feel proud.

It occupies an unique position, Rainy River itself forming a natural boundary between the District and the State of Minnesota; the Province of Manitoba bounding it on the west and north, and Thunder Bay District on the east.

The construction of the Canadian Pacific Railway, begun in 1877, revealed the possibilities of the Rainy River District as a home for the settler, and but a short time elapsed thereafter before the pioneers of a new land located on the shores of the Lake-of-the-Woods, Rainy Lake and Rainy River. Settlement will be greatly increased now that the attractions of the District are being made known and other railways are being constructed through the southern and south-western parts.

The extent of Rainy River, the quality of the soil in the District, the extent of the timber and mineral resources, the climatic advantages of the country, the nature of and market for the products of the soil, the prospects of the District in respect to settlement, and the many advantages open to the settler are dealt with in the paragraphs which follow.

The Act setting apart and forming the District of Rainy River, passed in 1885, will be found on page 23.

The chapter dealing with free grants and homesteads gives full information as to the manner in which these are to be obtained, including the Acts of the Ontario Legislature relating to the subject.

Under the head of mining, the principal clauses of the General Mining Act and recent amendments thereto are included, which provide for the sale of mining lands by the Crown Lands agents of the Province, instead of requiring the applicant to complete his purchase at the Crown Lands Department in Toronto as heretofore.

Another chapter includes the laws relating to the preservation of the forests from fire, ample machinery to that end being provided.

In conclusion, the affidavits called for in the Acts above referred to are inserted.

PART I.

THE RAINY RIVER DISTRICT:

ITS SOIL, CLIMATE, PRODUCTS, RAILWAY FACILITIES, SETTLEMENT, PROSPECTS AND TIMBER AND MINERAL RESOURCES.

OFFICIAL REPORT OF THE EXPLORATION OF RAINY RIVER DISTRICT.

During the summer of 1886, Mr. Thomas O. Bolger, of Belleville, a Provincial Land Surveyer, acting under instructions from the late Commissioner of Crown Lands (Hon. T. B. Pardee), proceeded to explore the land lying north of Rainy River and Rainy Lake. He reported as follows :

BELLEVILLE, ONTARIO, December 1st, 1886.

SIR,—I have the honour to report that in accordance with your instructions dated May 29th, 1886, I have explored the lands lying north of the surveyed townships on Rainy River, and also the country lying north of Rainy Lake.

I proceeded first to Rat Portage, where I procured the necessary supplies and canoes, and hired some men to assist in moving camp, etc., and then went across the Lake-of-the-Woods to the south shore of Sabashkong Bay, pitching my first camp at the mouth of Split Rock River, and from here explored the country south to the forty-ninth parallel, and eastward to the canoe route which leads from the easterly end of Sabashkong Bay to Fort Francis. I then followed the southerly shore of the Lake-of-the-Woods westward to the mouth of Rainy River, stretching inland sufficiently often to obtain a good general idea of the nature of the country and timber. I ascertained in this way that the land lying north of the forty-ninth parallel is generally of a very poor description, with the exception of some good patches in the vicinity of the Indian Reserve on Big Grassy River ; while the timber is generally poplar and jack-pine of small growth. I first encountered good land at the point where the forty-ninth parallel or the first base strikes the Lake-of-the-Woods, and following up Little Grassy River, which empties into the lake a couple of miles south of this point, I found, from travelling in every direction, that the block of four townships composed of townships one and two south, ranges twenty-three and twenty-four east, contains a large percentage of the finest land I have ever seen, and the same description applies to the block of land lying westward between these townships and the Lake-of-the-Woods. Little Grassy River is navigable for canoes for a distance of about eight miles from its mouth, and the land on the shore is all good, being composed of a rich calcareous drift formation, equal to any soil in the best agricultural districts of Ontario.

The timber along the river is chiefly large thrifty poplar, mixed with some scattering oak and swamp elm, and some evergreens such as balsam and spruce ; inland the timber changes in character somewhat from that along the river shore, as large Balm of Gilead, spruce, balsam and tamarac are met with more frequently, and the nice open bush which prevails along the river banks is changed for a tangled brushy undergrowth ; but the character of the soil remains the same. Tamarac and spruce swamps occur frequently in this section of the country, as is the case all through this large level area of good land which lies along the banks of Rainy River. These swamps were all perfectly dry this summer, and are nearly all capable of being made into excellent land by drainage, as they lie nearly as high as the surrounding dry lands, and only require proper ditching to take the surface water off in wet seasons. The extreme levelness of the country causes the presence of so   ch swamp land here, as the surface water has no means of escaping from the low-lying portions, and consequently the growth of moss and swamp timber is engendered. I noticed that in most cases the beds of the little streams are deep enough to form outlets for ditches and drains, and these creek beds are usually so numerous that to drain any swamp, no very long ditches would be required; in nearly all the swamps through which I passed I observed the soil to be a black vegetable mould, varying in depth from one to three feet, and always underlaid by the same calcareous clay above alluded to. I seldom met the muskeg proper, that is to say, the wet shaky bog in which water is present at all seasons of the year, and which grows nothing but dwarf spruce and moss. I then paddled up Rainy River, and on both shores I found the same kind of country a.. I have described as being in the vicinity of Grassy River, and as there are a good number of settlers along the river on the Canadian side I had an opportunity to observe the soil while under cultivation, and to see the kind of crops it is capable of raising.

The soil I found to be most excellent in character, calcareous clay overlaid by a thin streak of whitish fine earth about six inches in thickness, and this again covered with a coating of vegetable mould, and these three mixed up together in the working of the land form a soil which cannot be excelled in any part of the Dominion. I saw along the river crops of potatoes, turnips, hay, oats, wheat, corn, tomatoes and cabbage all grown to perfection this season, which shows that the climate, as well as the soil, is suitable to successful farming, especially when tomatoes ripen as they certainly did this year as well as ever I saw them ripen in the vicinity of Lake Ontario.

As I went up the river I frequently travelled inland several miles, and at the easterly side of township three, range twenty-four. I penetrated northward to the section I had explored from Grassy River, and I found that the calcareous clay formation extends at this point clear from the Lake-of-the-Woods to Rainy River, a distance of over twenty miles in a straight line; I found a tremendous bush fire raging along the first correction line south, which was destroying everything before it; in fact, bush fires were very frequent in this part of the country this season, owing to the extremely dry weather. There is an area of pine land in here a little north of the first correction line south, where the soil is inclined to be sandy, but the extent of this tract is not very large.

Along the line dividing ranges twenty-six and twenty-seven the good land extends back some twelve miles from the river, but towards the north-east corner of the township three the rough regions begin to appear, and away to the northward the country is broken and rocky, and the good land disappears.

Township three and the north part of township four, range twenty-seven, have been burnt over some years ago, and are now grown up with small second growth of poplar.

Townships four, in ranges twenty-eight, twenty-nine and thirty, are mostly all good land; while townships three, in the same ranges, are generally broken with rocky ridges, but contain some excellent land in the valleys among the hills; townships two, ranges twenty-seven and twenty-eight, also contain some good land, although broken by rocky hills.

A straight line drawn from the south-west corner of the large Indian Reserve on Big Grassy River to Fort Francis would approximately form the north boundary of the good belt of land, while almost all the country lying between this line and the Rainy River and the Lake-of-the-Woods is good agricultural land. This tract of country is over sixty miles long, and averages over fifteen miles wide, and contains over nine hundred square miles, or something like six hundred thousand acres, and has a water frontage on the Lake-of-the-Woods and Rainy River of over one hundred miles. Of this area perhaps thirty per cent. is swamp, most of which can be drained and made tillable land; rocky ridges occur very rarely, and the soil is all a limestone clay such as I have described. No limestone rock in place has been observed, but loose limestones containing. fossils are frequently to be met along the rivers, and the settlers along Rainy River pick them up and burn them into excellent lime; in fact, this whole district is a glacial drift.

The timber is chiefly poplar, which grows to a great size; I have seen trees over eighteen inches across the stump and sixty feet long clear of limbs. Balm of Gilead, too, prevails in some sections, while spruce, tamarac and balsam of thrifty growth are everywhere met with. In some places magnificent cedar abounds large enough for telegraph poles, shingle bolts, or any other use to which cedar is applied; there are some groves of pine through this section, but it cannot be called a pine country, that is, on the drift formatio .

North of the above imaginary line the country is all rough and broken with valleys of clay land occurring occasionally among the ridges, especially along the margins of creek beds; east of the line dividing ranges twenty-six and twenty-seven there is a good deal of pine, although in some places the fire has been through and destroyed much valuable timber; all around the north-west bay of Rainy Lake, and around the chain of waters stretching from this bay to the south-east corner of the Lake-of-the-Woods, I saw a considerable quantity of pine, both red and white, and in the country lying between this chain of lakes and the north bay of Rainy Lake, pine is present almost everywhere, but not often in large thick groves.

I explored all the country north of Rainy Lake, nearly as far north as the forty-ninth parallel, and eastward to what is called Sand Island River on the map, and up the Seine River to Sturgeon Falls; I travelled inland through this section of the country sufficiently often to get a good general idea of the land and timber. There is not much good land all through this region, that is, in large blocks, although patches of excellent clay land of from fifty to one hundred acres are met with frequently among the hills, but this clay is never the calcareous clay of the Rainy River drift. The only place I found a large tract of good land is on the bank of Sand Island River, extending from near the mouth of the river up stream for about sixteen miles, with a width of perhaps two miles; this tract is broken in places with rocky ridges, but this soil is a good clay loam and free from stone. This tract of good land has all been burnt over and is now grown over with small poplar. All the rest of this country, lying north of Rainy Lake, may be described as a rough, rocky region which in some places is utterly denuded of timber by forest fires. There is a considerable quantity of pine in all this section of country; all along the eastern shores of the north bay of Rainy River scattering pine is met with, and a good deal of lumbering has been done in the vicinity of the lake.

On the chain of water connecting Sand Island River with Rainy Lake, lumber camps have been in operation in former years. Around Sand Island Lake and in the country between this point and the River Seine there are some fine groves of red and white pine, and along the Seine also pine is frequently seen ; the other prevailing timber is chiefly jack-pine, with poplar and tamarac.

Along both sides of the Seine River and inland, both north and south, the country is rough and broken with occasional valleys of good land, and the same may be said of the land on Rat River, Pipestone River and Little Turtle River. From this it will be seen that the tract of country I have described adjoining Rainy River, and including the townships already surveyed, is a locality well adapted for farming, and although there is some good pine within this area it cannot be said to be a pine country.

The remainder of the country explored by me, including from Sabashkong Bay to Sturgeon Falls, on the Seine River, and north to the forty-ninth parallel, is comparatively unfit for settlement, but pine is met with all through this region, in some places only scattering, but in others in considerable groves, so that this portion may be classed as a lumbering district.

Wild rice is very abundant in all this country, and being an exceptionally good year for it the Indians laid in large quantities for winter use. Ducks, partridges and prairie chickens are very plentiful, and sturgeon, pickerel and whitefish are found in all the waters. Moose and caribou are very numerous, but the red deer are not found in these parts. Bears are very plentiful, but wolves are never seen in these woods.

While in the tent I wrote a detailed report of the result of each day's proceedings with the result of my explorations. I will send in a copy of this longer report at an early day, together with a map showing the routes taken by me each day.

I have the honour to be, Sir,
Your obedient servant,

(Signed),   THOMAS O. BOLGER,
Provincial Land Surveyor.

Soil.

As has been stated, Rainy River takes its course through a rich alluvial valley for over eighty miles. This valley is eminently adapted to support a large and populous agricultural population. As to the extent of the cultivatable land in the District, it is stated on good authority that all the land fronting on the river is suitable for settlement, while the arable area reaches back to a distance of sixteen miles, where the soil is found to be even superior to that at the water's edge, being mostly clay and clay loam with very little gravel or sand.

The greater part of the arable land requires no drainage whatever; even the occasional swamps of spruce, cedar and tamarac, are all dry in summer and can be made most excellent land by a little systematic drainage, and as they are all at a considerable height above the river level, proper ditching would remove the surface water in the spri. g. The beds of the small creeks and streams are deep enough to provide adequate outlets, ditches and drains. The area of good land along Rainy River, which is about eighty miles long by from ten to twelve miles wide, is remarkably free from stones and rocks.

Products.

The richness of the soil and the equable climate combine to produce a wide range of cereals and fruits. Hay, oats, and spring and fall wheat are successfully grown; the products of the garden include potatoes, onions, corn, carrots, turnips, tomatoes and cabbage, while the smaller fruits, such as raspberries, strawberries and plums, grow in abundance.

As an instance of the length of the season, oats sown as late as the middle of June have fully ripened. The settler who makes the above statement has lived on the bank of Rainy River for thirteen years and has never had a failure of crops. During the summer of 1889, his farm yielded seventy bushels of potatoes from one bushel of seed. A practical experience such as this illustrates most forcibly the fertility of the soil. The market for all a farmer can raise is at his door, the large number of lumbermen in the district being the purchasers. The following prices for grain and produce prevailed in the spring of 1890: Wheat, $1; barley, 75c.; oats, 50c.; peas, $1; potatoes, 75c.; turnips, 50c.; onions, $3 to $4; hay, per ton, $10 to $15; steamboat wood on river bank, $1.25 per cord; cows, $40 to $100 each; yoke oxen, $100 to $150; horses, $300 to $400 for good team; butter, 3oc. per lb.; eggs, 25c. per doz.; pork, 15c. to 18c. per lb.; beef, 12c. to 15c. per lb.

Schools and Churches.

There are good schools for white children at Fort Francis, Emo, Rat Portage, Keewatin and Norman. At Fort Francis Indian Reserve, at Kitchechokeyo Reserve, in the township of Woodyat, at the Indian Reserve in Barwick (Manitou Rapids) and at Roseberry (near the Long Sault Rapids) are Indian schools under able teachers. Religious privileges are as yet limited outside of Rat Portage and its immediate neighborhood, but the inflow of settlers will no doubt soon overcome this disadvantage.

Climate.

The climate of the district is similar to that in the region of Lake Ontario. Though the winter may be colder it is proportionately dryer and instead of the rain and slush of the eastern winter, excellent sleighing continues from December to March. The climate has been proven to be very healthful by the settlers, while, as is pointed out elsewhere, it is well suited to the growth of a wide range of cereals.

The name Rainy River (a perversion of René River) is really a mis-
nomer inasmuch as the rainfall is not excessive. One of the settlers
thus describes the climate : " The summer is generally moist, with clear
sunshine most of the time. The fall is beautiful, and November all
through about the best month of the year. The spring is similar—very
dry, with bright clear sun until about the first of June. The winters are
very dry. The snow is loose, yet it seldom drifts. I have not seen what
I would call a snowdrift yet. A log or hump on the level land shows all
through winter. Sometimes I think I will never require another
overcoat. I have so seldom worn one that apparently the one I have
will last my lifetime. On the very coldest days, except for a short time
at sunrise, we can chop in the woods the day through."

Navigation.

Rainy River being navigable for eighty miles, adequate means of
transportation over its waters was provided through the enterprise of
those who realized the great possibilities of this region, as soon as the tide
of settlement turned to its shores. Four steamers run regularly during
the season from Fort Francis on the southern side of Rainy Lake to Rat
Portage and Keewatin on the Canadian Pacific Railway—a distance of
180 miles. In addition two new boats will be placed on this route
during the coming summer of 1890. Round trips are made weekly.

There are in all thirty-five craft of different kinds, including steamers,
tugs and barges, in service in the District.

During the winter the mail is carried on sleighs, leaving Rat Portage
for Fort Francis on the first and fifteenth of every month, calling at the
Hudson Bay Company's Fort Louise, at the mouth of Rainy River, at
Hughes & Co.'s saw mill in the township of Atwood, at Rapid River post
office in the township of Worthington, at Rainy River post office in the
township of Morley, at Emo post office in the township of Lash, at Big
Forks post office in the township of Woodyatt, and at Isherwood post
office in the township of Roddick.

Fish and Game.

To the sportsman this part of Ontario is indeed a " happy hunting
ground." The waters abound in fish and wild fowl are very plentiful.
The moose are unfortunately becoming very scarce in the District, but black
bears are very numerous, while wolves are also scarce. Buffalo were seen by
early settlers near Rainy River but have since disappeared. The common brown bear and the more rare and beautiful silver fox are among
the denizens of North-western Ontario. Beaver abound in the streams
and creeks, while the otter, ermine and mink are plentiful. Partridge,
grouse and water flow! of all kinds are also extremely plentiful.

Railway Facilities.

For some years the Rainy River District has had the advantage of
the Canadian Pacific Railway, which runs through it diagonally, stopping at

the following stations : English River, Martin, Bonheur, Ignace, Raleigh, Toche, Wabigoon, Barclay, Eagle River, Vermillion Bay, Gilbert, Parrywood, Hawk Lake, Rossland, Rat Portage, Keewatin, Deception, Kalmar and Ingolt.

The Ontario and Rainy River Railway, eighty miles of which have been subsidized by the Legislative Assembly of Ontario at the rate of $3,000 per mile, is now in course of construction and will run through the southern portion of the District, ultimately connecting with the Port Arthur, Duluth and Western Railway. This new line will open up large tracts of good agricultural lands, forests of the finest timber such as pine, tamarac, spruce, cedar and poplar, as well as mineral lands already proven to be rich in gold, silver and iron, the development of which only awaits further railway facilities.

Settlement.

The great advantage of this District, especially with regard to water and wood, over a prairie country, have attracted a number of settlers from Northern Dakota. Many of the settlers have been on their lands since 1874, and have good houses and barns, large clearances, good fences, and well bred stock. The Ontario Government gives a Free Grant to every *bona fide* settler, head of a family, malo or female, 160 acres of land, and if he desires will sell him an additional 80 acres adjoining, at $1 per acre, payable in three years, while any or all of their sons over 18 years of age may have 120 acres free and may purchase 80 acres each at $1 an acre.

Up to the summer of 1889 there were about 175 *bona fide* settlers on farms, and the number will no doubt by this time exceed 200, the total population of the district, including Fort Francis, numbering nearly one thousand, not including the Ojibways and other Indians.

On February 13, 1889, a proclamation was issued by the Lieutenant-Governor-in-Council, bringing into force on February 18, 1889, the Act respecting Free Grants and Homesteads to actual settlers on Public Lands in the district of Rainy River, which was passed during the session of the Legislature held in 1886. [See Act on page 31]. Immediately following this proclamation, the Department of Crown Lands set apart twenty townships in the district as Free Grant Townships. They are situate on the Canadian bank of the Rainy River and contain the choicest and most fertile land to be found in the district, each township having a frontage on the river.

The surveys made in 1876 by the Dominion Government (on the one mile square section plan, the same as has been followed in the North-West), have been adopted by the Ontario Legislature, which legalized these surveys by the Act of 1886, and provided that any lands in the Rainy River District considered suitable for settlement and cultivation may, by Order in Council, be appropriated as Free Grants upon the terms specified.

Colonization Roads and Bridges.

Since the settlement of the Boundary question, the Ontario Government has adopted a liberal policy regarding the construction of Colonization roads and bridges in the district. About $30,000 has been expended during the last five years on the three leading highways of the District alone, viz: the Rat Portage and Keewatin Road, the Rabbit Mountain and White Fish Lake Road, and the Rainy River Road. The first is about five miles in length, and serves the towns of Rat Portage, Keewatin, and neighboring localities; the second is about three miles long and leads to the great silver mining district in the neighborhood of and beyond Rabbit Mountain; while the third is some seventy miles in length and follows the course of Rainy River on the Canadian bank from Fort Francis to the Lake-of-the-Woods.

As to bridges, the Rat Portage and Keewatin Road necessitated the building of three large bridges across the main outlets of the Winnipeg River. Two large bridges have also been built across the Kaministiquia River to develop both the mining and farming interests.

The Timber Resources of the District.

The most valuable resource in Rainy River district is its timber, extending along the entire length of Rainy River, of pine, poplar, birch, basswood, oak, elm, ash, soft maple, balm of gilead, balsam, spruce, cedar, and tamarac.

Lumbering operations are carried on to a considerable extent on Rainy Lake and its eastern tributaries. There are also two well equipped sawmills on Rainy River where the incoming settler may procure the necessary material for the erection of a home, and where he has the additional advantage of obtaining employment during the winter months at wages ranging from $20 to $30 per month, and from $2 to $2.50 per day with board for team and teamster.

On the banks of the Seine and other rivers flowing into Rainy Lake, there is a very large growth of both red and white pine.

The Dominion Public Works report of 1875, in dealing with the pine-growing capacities of this region, says that extensive groves of red and white pine are to be found, of a size and quality well adapted to all the purposes for which such timber is usually applied. On the alluvial belt of Rainy River white pine of a large size is to be seen interspersed with other descriptions of forest trees, and on the Lake-of-the-Woods and mainland to the north and east there are occasionally pine groves of moderate extent, which lessens in quantity as it nears Lake Winnipeg where the pine belt finally disappears.

Lieutenant-Colonel Dennis, the late Deputy Minister of the Interior, has estimated the quantity of pine to be found between Lake Superior and the Lake-of-the-Woods—including that on the islands in the lake—at twenty-six thousand millions of feet, board measure.

The Manitoba *Colonist* for May, 1890, says :—" The following lumber mills are located on the Lake-of-the-Woods : At Rat Portage, Ross, Hall & Brown and the Western Lumber Co.; at Norman, Hy. Bulmer's, Cameron & Kennedy and the Minnesota & Ontario Lumber Co.'s; at Keewatin, The Keewatin Lumber & Manufacturing Co., and Dick & Banning's. These mills give employment to nearly 1,000 men. The lumber cut during the present season will be unprecedented in this District. Some 40,000,000 of Canadian logs as well as about 25,000,000 of American ones will be cut between the above mills. The reason for the increased activity is attributed to the fact that heretofore the lumbermen here had no *bona fide* right to cut the timber, but now the Ontario Government, having control, have given the lumbermen what they required."

THE

TOWNS AND VILLAGES IN RAINY RIVER DISTRICT.

Rat Portage.

The largest town on the Lake-of-the-Woods is Rat Portage, with its population of some 1,500. It is the seat of much business enterprise and from prospects is destined to grow. Being one of the divisional points of the Canadian Pacific Railway adds materially to its activity, while the fact that it is the principal port on the Lake-of-the-Woods also conduces to its prosperity. It has a well conducted weekly newspaper, the Rat Portage *News*, which is devoted to the interests of the whole District. It is also the judicial seat of the District, with the offices of the Stipendiary Magistrate, Sheriff, and other officers. Extensive fisheries do a good business, shipping the products of the Lake to Chicago, Minneapolis, Buffalo and Denver. It is also the principal shipping port of the District for lumber, etc.

Keewatin.

This town on the Lake-of-the-Woods has long been known through the western country as the seat of extensive lumber operations, and when the Lake-of-the-Woods Milling Co's Flour Mill was built with its accessories in the way of barrel factory and elevators, and the output of flour scattered far and wide over Canada and across the sea, the name Keewatin has become as familiar a household word as Rat Portage. The town possesses 4 general stores, 2 hardware stores, 2 bakeries, 2 groceries, 1 drug store, 2 butchers, 2 saw mills, 2 planing mills, a board factory, the largest flour mill in Canada, a brick yard, 2 hotels, a liquor store, a tailor and a doctor. The Presbyterians, Methodists and Roman Catholics all have handsome churches of their own here. The Foresters, Masons and Royal Templars of Temperance have societies in active operation. The magnificent new school is under able management. Population of the place is about 700. Numerous fine private residences have been built here. As a summer resort, Keewatin and vicinity claims its numbers each season.

Norman.

Norman, lying in the centre of the group of towns on the Lake-of-the Woods, is situate midway between Keewatin and Rat Portage. The water power here is unlimited, and there are excellent openings for factories. The following are the business enterprises already established: 4 general stores, 1 grocer and confectioner, 3 blacksmith shops, 3 saw mills, 3 planing mills, a machine shop, 1 licensed hotel and 7 boarding houses. There is

2 (R. R.)

a large public hall and services are held here regularly by the Presbyterians. The town has a population bordering on 800. As a summer resort Norman is one of the places of interest to tourists, many finding it a central point from which to purchase their supplies during their outing on the Lake-of-the-Woods.

Alberton (or Fort Frances).

The principal settlement on Rainy River is Fort Frances or Alberton. It has a school and church, and several general stores, saw mill, etc., and is destined to be an important centre of population.

Fort Frances Lock.

The works at Fort Frances consist of a canal 800 feet in length, cut through the solid rock, about 40 feet wide, with one lift of 24 feet 8 inches. The chamber of the lock is 200 feet long and 38 feet wide in the clear. The lowest depth of water on the sills will be 5 feet 6 inches, but it is rarely if ever known to be so low as that, and is ordinarily from 8 to 10 feet. The cost of the works to the Dominion Government has been $250,000.

A new Town (Lockington).

Mr. Alexander Locking, of Emo, one of the most successful settlers in the District, has surveyed a village site on the Canadian bank of Rainy River, the name of "Lockington" being registered. The government road runs through it, and will be named "Mowat Street."

VARIOUS DESCRIPTIONS OF RAINY RIVER DISTRICT.

Professor Macoun's Testimony.

Professor Macoun, who visited and reported upon this district in 1874, describes it as follows :—" The approach to Fort Frances is very beautiful. As we approach the outlet to the lake and enter Rainy River, the right bank appears very much like a gentleman's park, the trees standing far apart and having the rounded tops of those seen in open grounds. Blue oak (*Quercus Prinos var. discolor*), and Balsam Poplar (*Populus balsamifera*), with a few aspen, are the principal forest trees. These line the bank, and, for two miles after leaving the lake, we glide down between walls of living green until we reach the Fort, which is beautifully situated on the right bank of Rainy River, immediately below the falls. All sorts of grain can be raised here, as well as all kinds of garden vegetables ; little attention is given to agriculture, but enough was seen to show that nature would do her part if properly assisted. Barley three feet high, and oats over that, showed there was nothing in the climate or soil to prevent a luxuriant growth. * * * The length of the river is about eighty miles. The right, or Canadian bank, for the whole distance is covered with a heavy growth of forest trees, shrubs, climbing vines and beautiful flowers. The Indians say the timber gets larger as you proceed inland. The forest trees consist of oak, elm, ash, birch, basswood, balsam, spruce, aspen, balsam poplar, and white and red pine near the Lake-of-the-Woods. The whole flora of this region indicates a climate very like that of central Canada, and the luxuriance of the vegetation shows that the soil is of the very best quality. Wild peas and vetches were in the greatest profusion ; the average height was about six feet, but many specimens were obtained of eight feet and upwards. While the boat was wooding I took a stroll inland, and found progress almost impossible owing to the astonishing growth of herbaceous plants. The following plants were observed on Rainy River, and are only an index to the vast profusion of nature's bounties in that region :—Lilium Canadense, Lilium Philadelphicum, Vicia Americana, Calystegia spithamea, Calystegia sepium, Aralia hispida, Lobelia Kalmii, Smilacina stellata, Lathyrus venosus, Lathyrus ochrolencus, Monarda fistulosa, Viburnum pubescens, Astragalus Canadensis, Erysimum chieranthoides, Asarum Canadensis, and Lopaulthus anistatus."

Sir George Simpson's Description.

Writing of the Rainy Lake region, Sir George Simpson, in his "Overland Journey Round the World" (1841-2, page 45), was fully as eulogistic of its merits and beauties as he had been of those of the Kaministiquia valley. His description agrees remarkably with that of Mr. Macoun just quoted. Sir George Simpson says :—" From Fort Francis downwards, a stretch of nearly 100 miles, the river is not interrupted by a single impediment, while yet the current is not strong enough to retard an ascending traveller. Nor are the banks less favourable to agriculture than the waters themselves to navigation, resembling, in some measure, those of the Thames, near Richmond. From the very brink of the river there rises a gentle slope of green sward, crowned in many places with a plentiful growth of birch, poplar, beech, elm and oak. Is it too much for the eye of philanthropy to discern through the vista of futurity this noble stream, connecting as it does, the fertile shores of two spacious lakes, with crowded steamboats on its bosom and populous towns on its borders ?" A few years later, before a Select Committee of the House of Commons in London, Sir George endeavoured to qualify, to some extent, his former glowing panegyric. But he was at that time looking on this and some other matters in question, not with " the eye of philanthropy," but through a pair of Hudson's Bay monopoly spectacles, and, under a vigorous cross-examination by Mr. Roebuck, had virtually to admit the correctness of his first description, founded, as it was, on an experience of of twenty-seven years.

Mr. S. J. Dawson's Description.

The report of Mr. S. J. Dawson, in 1874 (then engineer in charge of the district), fully corroborates the views of the two eminent authorities already quoted. He says :—" Alluvial land of the best description extends along the banks of Rainy River in an unbroken stretch of seventy-five or eighty miles from Rainy Lake to the Lake-of-the-Woods. In this tract, where it borders on the river, there is not an acre unsusceptible to cultivation. At intervals there are old park-like Indian clearings, partly overspread with oak and elm, which, although they have naturally sprung up, have the appearance of ornamental plantations. *
* * The whole district is covered with forests, and Canadian settlers would find themselves in a country similar, in many respects, to the land of their nativity ; nor does the climate differ essentially from that of the most favoured parts of Ontario or Quebec. Wheat was successfully grown for many years at Fort Frances, both by the old North-west Company and their successors, the Hudson's Bay Company. The Indians still cultivate maize on the little farms on Rainy River and Lake-of-the-Woods. In many places the wild grape grows in extraordinary profusion, yielding fruit which comes to perfection in the fall. Wild rice, which requires a high summer temperature, is abundant, and, indeed the flora, taken generally, indicates a climate in every way well adapted to the growth of cereals."

PART II.

THE ACT SETTING APART AND FORMING

THE

DISTRICT OF RAINY RIVER.

AN ACT RESPECTING THE DISTRICT OF RAINY RIVER.

[The following is the Act setting apart the District of Rainy River as a territorial district which was assented to 30th March, 1885.]

CHAPTER 20, STATUTES OF ONTARIO, 1885.

WHEREAS the Lieutenant-Governor in Council, on the *Preamble.* third day of October, in the year of our Lord 1884, by virtue of an Act passed by the Legislature of Ontario, in the session thereof held in the 47th year of Her Majesty's reign, entitled *An Act respecting the District of Algoma and Thunder Bay*, issued a proclamation naming the 11th day of the said month of October, as the day upon which the said Act respecting the District of Algoma and Thunder Bay should go into force; and whereas the Lieutenant-Governor in Council, on the 13th day of January, in the year of our Lord 1885, in pursuance of the powers in the said Act contained, did proclaim and declare that, from and after the 15th day of February then next, all that part of the Provisional Judicial District of Thunder Bay lying west of a line drawn due north and south through the most easterly point of Hunter's Island should, for the purposes (except registry purposes) mentioned in the Revised Statute *respecting the Territorial Districts of Muskoka, Parry Sound and Thunder Bay*, be detached from the said Provisional Judicial District of Thunder Bay, and should form a separate Territorial District by the name of The District of Rainy River; and whereas it is expedient to make provision in respect of the matters hereinafter mentioned:

Therefore Her Majesty, by and with the advice and consent of the Legislative Assembly of the Province of Ontario, enacts as follows:—

1. The said Territorial District of Rainy River, being all *District of* that portion of the Province lying west of the said line, shall, *Rainy River separated* from and after the first day of July next, also be separated, *from District* for registry purposes, from the District of Thunder Bay, and *of Thunder Bay for registration purposes.* shall form a separate registry division.

2. (1) The Lieutenant-Governor may, from time to time *Appointment of Deputy* appoint, under the great seal, an officer for the District Court *Clerk.* of the Provisional Judicial District of Thunder Bay, to be called the Deputy Clerk for Rainy River, who shall hold office during pleasure, and shall keep his office at Rat Portage.

(2) In case after an appointment has been made a vacancy *Vacancy in* occurs in such office, the Clerk of the Division Court at Rat *the office of Deputy Clerk.* Portage shall, *ex-officio*, be Deputy Clerk until another appointment is made.

Powers and Duties of Deputy Clerk. (3) The said Deputy Clerk shall issue writs for the commencement of actions in the said District Court, and shall, in respect of actions so commenced and of proceedings therein, perform the like duties and have the like powers and rights as are performed or possessed by the Clerk of the District Court at Port Arthur in respect of actions commenced by writs issued out of his office, and of proceedings therein ; and the said Deputy Clerk shall also issue such writs and process as may be required in such actions as may in like cases be issued by the said Clerk of the District Court, and may renew any such writs as by law may be renewed.

Capias. (4) No writ of capias issued under the next preceding subsection shall be executed outside of the District of Rainy River ; and every writ of capias so issued shall be marked by the Clerk as follows : ' Only to be executed within the District of Rainy River," but this shall not prevent a copy of such writ of capias being served at any place within Ontario.

Seal. (5) The Deputy Clerk of the said District Court shall have the custody of a seal similar in design to the seal of the court in the custody of the Clerk at Port Arthur, and the said Deputy Clerk shall seal with the said seal all writs, process and proceedings requiring the seal of the said court ; and every writ, process or proceeding sealed with such seal shall be held to be duly sealed with the seal of the said court.

Venue. 3. In any actions in which the venue is local the writ shall be issued out of the office of the said Deputy Clerk, and the venue shall be laid in the District of Rainy River in the same manner as if the said district was a separate county ; but the judge may, if he sees fit, change the venue in any action.

Deputy Clerk to be Registrar of Surrogate Court. 4. (1) The Deputy Clerk for the Rainy River District of the District Court of Thunder Bay shall, *ex-officio*, be Deputy Registrar for Rainy River of the Surrogate Court of Thunder Bay ; and he shall keep his office of Deputy Registrar at the same place as he is required by law to keep his office of Deputy Clerk.

R. S. O., c. 46, ss. 10-13, to apply to Deputy Registrar. (2) Sections 10, 11, 12 and 13 of the Revised Statute, chapter 46 (*The Surrogate Courts Act*), shall apply as nearly as may be to the Deputy Registrar for Rainy River ; and he shall observe and conform to the provisions thereof ; and shall perform the like duties, and shall have the like powers and rights, under and by virtue of the said Revised Statute, within the District of Rainy River, as are performed or possessed by the Registrar of the Surrogate Court for Thunder Bay at Port Arthur ; and the latter shall, after the passing of this Act, cease to exercise the powers and rights of Registrar of the Surrogate Court for Thunder Bay, in regard to applications for probate, or letters of administration, in respect of the will, or estate, of any person who had at the time of his death his

25

fixed place of abode in the District of Rainy River, or of any person who having no fixed place of abode within Ontario had, at the time of his death, real or personal estate in such District, which but for this Act would have been exercised by him as Registrar of the Surrogate Court for Thunder Bay.

(3) The said Deputy Registrar of Surrogate shall have the custody of a seal similar in design to the seal of the court in the custody of the Registrar, and such seal shall be the seal of the court for the purpose of sealing all grants, letters, writs, certificates, papers or proceedings in connection with any matter or thing in the office of the said Deputy Registrar requiring to be sealed. *Surrogate Seal.*

5. The Surrogate Court for Thunder Bay shall, at Rat Portage in the District of Rainy River, in respect of matters arising within the District of Rainy River, and at Port Arthur in respect of matters arising within the rest of the Provisional Judicial District of Thunder Bay, hold such sittings as the Judge of the Surrogate Court of the Provisional Judicial District of Thunder Bay may think proper and necessary, but the said Judge may, when he deems it more convenient for the parties interested, perform any judicial or magisterial act affecting either of the said Surrogate divisions in the other of such divisions. *Sittings of Surrogate Court.*

6.—(1) The Lieutenant-Governor may also appoint a Sheriff of the said District of Rainy River, who shall keep his office at Rat Portage. *Appointment of Sheriff.*

(2) All writs and other process requiring to be directed to a Sheriff and intended to be executed within the said District of Rainy River shall be directed to the said Sheriff.

(3) Nothing herein contained shall prevent the Sheriff of Thunder Bay from proceeding upon and completing the execution or service within the said District of Rainy River, of any writ of *mesne* or final process in his hands at the time this Act takes effect, or any renewal thereof, or any subsequent or supplementary writ in the same cause; or in the case of executions against lands, from executing all necessary deeds and conveyances relating to the same; and the acts of the said Sheriff of Thunder Bay in respect of these matters shall be valid in the same manner and to the same extent as if this Act had not been passed, and no further.

(4) Sub-sections 5, 6, 7 and 8 of section 12 of the Act passed in the 43rd year of the reign of Her Majesty, entitled, *An Act respecting the Administration of Justice in the District of Algoma' Thunder Bay and Nipissing*, shall apply to the District of Rainy River and to the Sheriff thereof.

7. Unless where inconsistent with this Act and as nearly as may be, the Acts mentioned in schedule A appended to this Act shall, to the extent shewn in the third column of the said *Application of certain Acts to District of Rainy River.*

schedule, apply to the District of Rainy River, and all other . Acts, or parts of Acts, applying in general terms to Territorial Districts, shall also apply to the said district.

Returns of Convictions.

8. All returns of convictions required by law to be made by any Justice or Justices of the Peace shall, for the District of Rainy River, be made to the Clerk of the Peace of the District of Thunder Bay.

Sitting of District Court.

9.—(1) Besides the sittings at the district town, the District Court of Thunder Bay shall hold sittings on the first Tuesday of the month of June and the fourth Tuesday of the month of November of each year, at Rat Portage, for trials and assess-ments by jury in cases in which the venue is laid in Rainy River, and sittings of the General Sessions of the Peace of Thunder Bay shall be held on the same days.

Sitting of General Sessions.

Trial of appeals.

(2) The said General Sessions of the Peace shall be for the trial of causes within the jurisdictions of the General Sessions where the offence to be tried was committed within the District of Rainy River, and for the trial of appeals to the General Sessions from a decision, order or conviction made by a Justice of the Peace within such district.

Gaols and Lock-ups.

10.—(1) Any gaol or lock-up erected in the said District of Rainy River under the authority of the Lieutenant-Governor, or any building so declared by Order in Council, shall be a common gaol of such district, for the safe custody of persons charged with the commission, within the said District, of crimes, or with the commission therein of offences against any statute of this Province, or against any municipal by-law, who may not have been finally committed for trial, or for the safe custody of persons finally committed for trial, charged as aforesaid, who are to be tried within the said District of Rainy River ; or for the confinement of persons sentenced within the said district for crimes or for offences as aforesaid, for periods not exceeding six months ; or for the confinement of persons sentenced as aforesaid for periods exceeding six months until such persons can be conveniently removed to the gaol at Rat Portage, or other lawful prison to which they are sentenced.

Gaol at Rat Portage.

(2) The gaol at Rat Portage shall be the chief common gaol of the District, and, besides being for the detention of persons held for safe custody as mentioned in this section, shall also be for the confinement of persons sentenced within the said District for crimes or offences as aforesaid for periods less than two years. .

Division Courts.

11. The Third and Fourth Division Courts of the District of Thunder Bay, the limits of which are now within the District of Rainy River, shall, after the first day of April, 1885, be respectively known as the First and Second Division Courts of

the District of Rainy River, subject to the authority of the Lieutenant-Governor in Council to alter the numbers, limits and extent of the divisions.

12. Whereas the dispute with respect to the Boundary between this Province and the Province of Manitoba has been determined in the manner contemplated by the Act passed at the last session of the Ontario Legislature, chapter two, entitled *An Act respecting the territory in dispute between this Province and the Province of Manitoba,* the said Act is therefore hereby repealed except the 27th, 28th and 29th sections thereof; and whereas the report in that behalf of the Judicial Committee of the Privy Council bears date the 22nd day of July last, and the Order of Her Majesty in Council confirming the same bears date the 11th August following, but the determination of the dispute was not immediately known in the disputed territory, it is hereby declared and enacted that the said Act shall be deemed to have been in force notwithstanding anything therein contained, until the 26th day of October last, but no longer; and the authority of the council at Rat Portage which was suspended by the said Act is hereby declared to have been revived from the 26th day of October aforesaid, and the by-laws, rules and regulations theretofore passed or enacted by the Municipal Board of Rat Portage shall, except so far as they have been since varied by the said council, be held to be as valid and effectual as the same would have been had the authority of the said council not been suspended, and had such by-laws, rules and regulations been passed or enacted by the said council.

47 V., c. 2, repealed except as to sections 27, 28, 29.

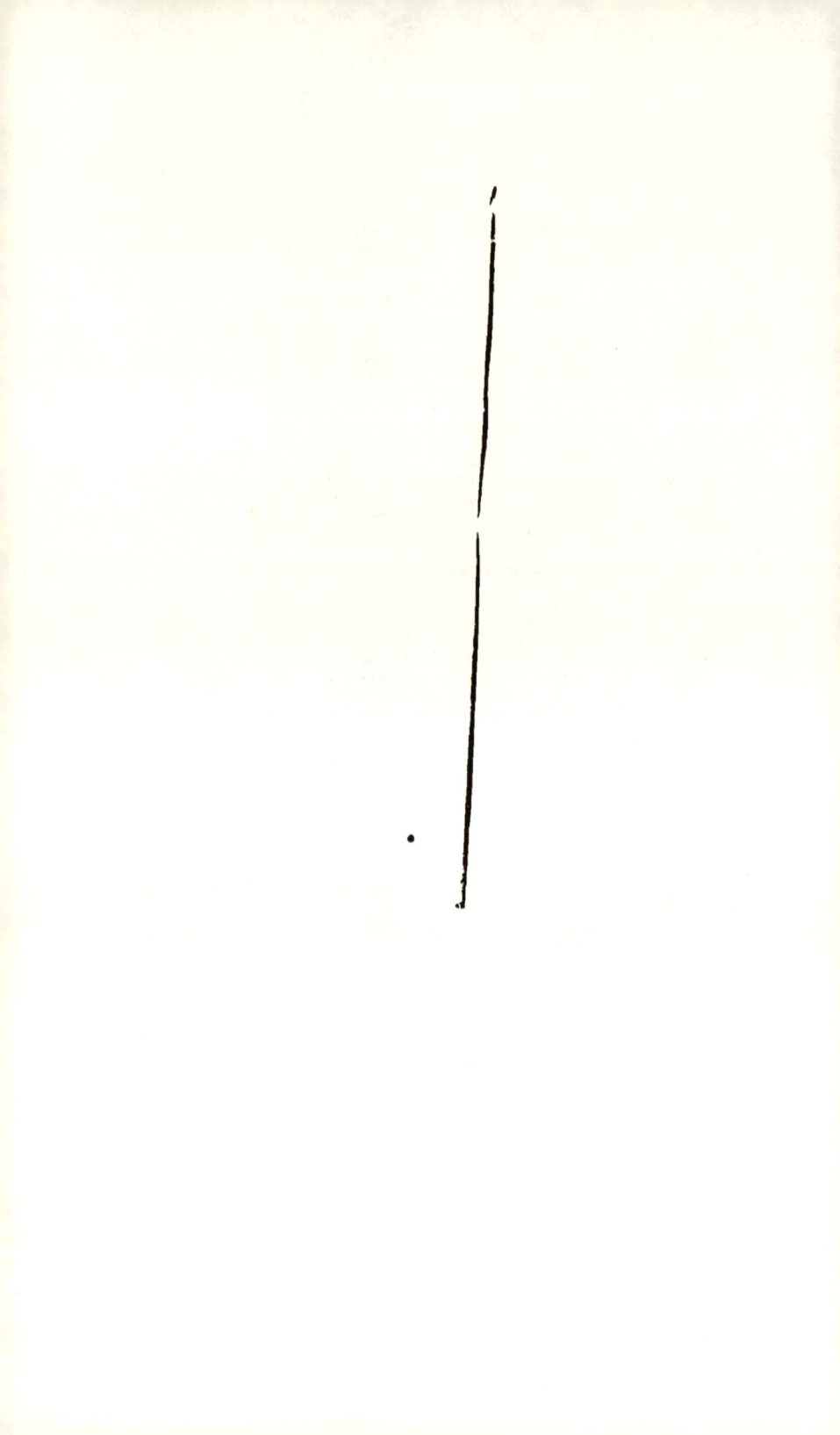

PART III.

THE RAINY RIVER FREE GRANTS ACT.

LIST OF FREE GRANT TOWNSHIPS.

DIRECTIONS AS TO HOW TO OBTAIN FREE GRANTS AND PUBLIC LANDS.

THE RAINY RIVER FREE GRANTS AND HOMESTEADS ACT.

CHAPTER 26, REVISED STATUTES OF ONTARIO, 1887.

An Act respecting Free Grants and Homesteads to Actual Settlers on Public Lands in the District of Rainy River.

WHEREAS under instructions from the Department of the Interior of Canada, certain townships have been surveyed in the Rainy River District, the lots immediately upon the bank of Rainy River having a width of ten chains fronting the river and a varying depth, and the remaining land so surveyed being subdivided into sections of one mile square, and quarter sections of one hundred and sixty acres, with road allowances around each section; and whereas a number of settlers have gone into occupation of the lands so surveyed, and it is expedient to adopt said surveys and otherwise provide for the settlement of the lands in question ; *Preamble.*

Therefore Her Majesty, by and with the advice and consent of the Legislative Assembly of the Province of Ontario, enacts as follows :—

1. This Act may be cited as " *The Rainy River Free Grants and Homesteads Act.*" 49 V. c. 7, s. 1. *Short title.*

2. The said surveys are hereby adopted and legalized, and the Department of Crown Lands is authorized to continue such system of survey within the District of Rainy River, so far as may be deemed expedient. 49 V. c. 7, s. 2. *Former surveys adopted.*

3. The Lieutenant-Governor in Council may appropriate any lands in the Rainy River District considered suitable for settlement and cultivation, and not being mineral lands or pine timber lands, as free grants to actual settlers, under such regulations as shall from time to time be made by Order in Council not inconsistent with the provisions of this Act. 49 V. c. 7, s. 3. *Appropriation of lands for settlement.*

4. *The Free Grants and Homesteads Act,* saving and excepting as is hereafter provided, and so far as the same is not inconsistent with the provisions of this Act, shall apply to lands opened for settlement under this Act. *Application of R. S. O. c. 25, to this Act.*

1. The male, or sole female, head of a family with children under eighteen years of age residing with him, or her, may be located for a free grant to the extent of one hundred and sixty acres, or a quarter section. *Free grants to heads of families.*

Free grants to males 18 years of age.

2. A male of the age of eighteen years, without children may be located for a free grant to the extent of one hundred and twenty acres, or a half quarter section, together with an adjoining quarter quarter section.

Purchase of locations for children.

3. In addition to location every head of a family having children under eighteen years of age residing with him, or her, may purchase at the time of location an adjoining half quarter section, or eighty acres, at $1 per acre, payable one-fourth cash and the balance in three equal annual instalments with interest.

Purchase of locations by males 18 years of age.

4. A male of the age of eighteen years, without children, entitled to locate, may purchase at the time of location an adjoining half quarter section, or eighty acres, at $1 per acre, payable one-fourth cash, and the balance in three equal annual instalments with interest.

Issue of patents.

5. Patents for lands located and purchased under this Act may issue at the expiration of three years from the date of location and purchase.

Sale to person who has made improvements.

6. Where a person has, previous to the passing of this Act, made substantial improvements on two or more adjoining lots, and the lots contain more land than the person is entitled under this Act to locate and purchase, the Commissioner of Crown Lands may sell to such person such additional quantity of land at $1 per acre as may, under the circumstances, seem just and equitable.

Issue of patents to persons having made improvements.

7. In case a person has occupied and made the required improvements upon one or more lots of land before the passing of this Act, the Commissioner of Crown Lands may, after location and purchase as hereinbefore provided, issue the patent therefor without waiting for the expiration of three years.

Reservation of pine trees, mines and minerals.

8 Pine trees growing or being upon any lands located or purchased under this Act, and gold, silver, copper, lead, iron or other mines, or minerals, shall be considered as reserved from the location or purchase, and shall be the property of Her Majesty, except that the locatee, or those claiming under him, may cut and use such trees as may be necessary for the purpose of building, fencing and fuel, on the land so located or purchased, and may also cut and dispose of all trees required to be removed, in actually clearing the land for cultivation, but no pine trees (except for the necessary building, fencing, and fuel as aforesaid) shall be cut beyond the limit of the actual clearing before the issuing of the patent; and pine trees so cut and disposed of (except for the necessary building, fencing and fuel as aforesaid) shall be subject to the payment of the same dues as are at the time payable by the holders of licenses to cut timber or saw logs.

Trees to pass to patentee.

9. Trees remaining on the land at the time the patent issues shall pass to the patentee. 49 V. c 7, s. 4.

Commencement of Act.

5. This Act shall not go into force until a day to be named by the Lieutenant-Governor by his proclamation. 49 V. c. 7, s. 5.

AN ACT TO AMEND THE FREE GRANTS AND HOMESTEADS ACT.

The following amendments to the Free Grants and Home-
steads Act, and which apply also to the Rainy River
Free Grants and Homesteads Act, was passed by the
Ontario Legislature during the Session of 1890 :

HER MAJESTY, by and with the advice and consent of the
Legislative Assembly of the Province of Ontario, enacts
as follows :—

1. Section 12 of *The Free Grants and Homesteads Act* is
repealed, and the following is substituted therefor :— Rev. Stat. c. 25, s. 12, repealed.

12. From and after the 30th day of April, 1880, the patentee,
his heirs or assigns, of land located or sold under this Act, after
the 5th day of March, 1880, shall be entitled to be paid out of
the Consolidated Revenue of the province, on all pine trees cut
on such land subsequent to the 13th day of April next, after
the date of the patent, and upon which dues have been collected by the Crown, the sum of thirty-three cents on each one
thousand feet, board measure, of saw logs, and four dollars on
each one thousand cubic feet of square or wancy timber, and
the Lieutenant-Governor in Council may make regulations for
ascertaining and determining the persons from time to time to
receive the payments and the sums to be paid. Payment by Crown to patentees of part of dues.

2. In case a person who has complied with all the settlement duties under the said Act and obtained a patent for only
one lot, is entitled to and desires to obtain another 100
acres to make up his full quantity, or having obtained his full
quantity as a free grant has purchased an additional 100 acres
under the orders and regulations under the said Act, and such
additional location or purchase is adjacent to his patented lot,
the Commissioner of Crown Lands upon being satisfied that
such lot or lots are not chiefly valuable for their pine timber,
and are suited only or principally for grazing purposes or as a
fuel reserve, may dispense with residence and settlement duties
upon them provided there are 30 acres cleared upon the
patented lot, and may issue the patent at the expiration of the
time required by the said Act. Commissioner may dispense with residence and settlement duties in certain cases.

3. In case a person is *bona fide* the owner and occupant of
land in a free grant district acquired otherwise than as a free
grant under the said Act and is entitled and desires to obtain
a free grant location and such location is adjacent to the land
which he owns and occupies, the Commissioner of Crown
Lands upon being satisfied by inspection or evidence that the
lands are not chiefly valuable for their pine timber, and are
suited only or principally for grazing purposes or as a fuel
reserve, and that there are 30 acres cleared upon the land Commissioner may dispense with clearing and residence in certain cases.

3 (R. R.)

which he owns and resides upon, may dispense with clearing and residence upon such free grant location and issue the patent at the expiration of the time required by the said Act.

Where occupant for six years not regularly located through inadvertence patent may issue before five years.

4. In case a person entitled to obtain a location under the provisions of the said *Free Grants and Homesteads Act* has, without objection by the Crown for a period of six or more years occupied or made the required improvements upon one or more lots (not exceeding the quantity which may be granted under the said Act) of land prior to the said land being brought under the operation of the said Act, or if the land be open for location, in case he has so occupied but has not, either through inadvertence or oversight, been regularly located, the Commissioner of Crown Lands, subject to such regulations as may be provided in that behalf, may after location as by said Act is provided, issue the patent upon proof of the performance of the required settlement duties and without awaiting for the expiration of five years from the date of location. But this section shall not apply where it appears to the Commissioner that the lot has been selected chiefly on account of the pine timber thereon.

Rev. Stat. c. 25, s. 10, amended.

5 Section 10 of *The Free Grants and Homesteads Act* is amended by adding thereto the following sub-section :—

Locatee of two or more lots may cut pine for building and fencing.

(2) Where the land allotted to a locatee or purchaser under this Act, is composed of two or more lots, or parcels of lots, the said locatee or purchaser, or those claiming under him may cut such pine trees as may be necessary for the purpose of building and fencing as hereinbefore provided, or any one or more of the said lots or parcels of lots so located or purchased, and may use the said pine trees on the same lot or any of the other lots or parcels of lots held by him as a free grant or by purchase under this Act, whether located at the same time or otherwise.

Rev. Stat. c. 25, s. 22, sub-ss. 3 and 4 repealed.

6. Sub-sections 3 and 4 of section 22 of *The Free Grants and Homesteads Act*, are hereby repealed and the patents for the lands in said sub-sections mentioned or referred to may issue notwithstanding any arrears of payments of the expenses of clearing, fencing and erection of buildings thereon, and all

Remission of dues from settlers in Ryerson and Spence.

sums due Her Majesty in respect of such clearing, fencing and building by locatees in the townships of Ryerson and Spence, in the district of Parry Sound, amounting to \$7,304 principal, together with any interest thereon, are hereby remitted.

Sums due to Crown for seed grain remitted.

7. That all sums due to the Crown for seed-grain supplied to the settlers in the free grant territory or any part thereof, amounting to \$3,306, together with any interest thereon, are hereby remitted.

Act incorporated with Rev. Stat, c. 25.

8. This Act shall be read with and as part of *The Free Grants and Homesteads Act.*

Rainy River Free Grant Townships.

The following townships are now open for location under the Rainy
River Free Grants and Homesteads Act :—

Township of—

Curran	Township 3	Range 22
Atwood	" 4	" 22
Blue	" 3	" 23
Worthington	" 4	" 23
Nelles	" 3	" 24
Dilke	" 4	" 24
Patullo	" 3	" 25
Morley	" 4	" 25
S. of Morley.............	" 5	" 25
Tait	" 3	" 26
Shenston.................	" 4	" 26
Roseberry	" 5	" 26
Barwick	" 5	" 27
Lash	" 5	" 28
Aylsworth	" 6	" 28
Devlin	" 5	" 29
Woodyatt	" 6	" 29
Crozier..................	" 5	" 30
Roddick	" 6	" 30
McIrvine	" 5	" 31

How to Obtain Free Grants and Homesteads in the Rainy River District.

Public lands which have been surveyed, and are considered suitable
for settlement and cultivation, and not valuable chiefly for minerals or
pine timber, may be appropriated as Free Grants.

To obtain a Free Grant, the applicant must make application to the
local Crown Land agent, in whose agency the land desired is situated,
and deposit with him the necessary affidavit (see Forms Nos. 1, 2 and 3,
in Appendix). Although no fees are charged by the Department, or
allowed to the land agents for locating, yet if required to prepare the
necessary affidavits, the agent may make a reasonable charge for so doing.

One hundred and sixty acres is the limit of the Rainy River Free
Grants and Homesteads Act. No individual, therefore, can obtain more
than that quantity as a Free Grant, and if the land selected exceeds 160
acres, the applicant must pay for the overplus at the price fixed by
the regulations, viz., $1 per acre. *The male head of a family,
or the sole female head of a family, having a child or children under*

eighteen years of age residing with him or her, may be located for 160'
acres as a Free Grant; and may also purchase an additional 80 acres
at the rate of one dollar per acre.

Upon receipt of the necessary affidavits, the agent will, if the land
selected be open for location, and there be no adverse claim thereto, enter
the locatee for it on the records of his office, and at the end of the current
month he will return the location to the Department of Crown Lands.

In case a party has settled on Government land before the township
has been surveyed, or appropriated under the Free Grants Act, he should,
immediately after it is opened for location, apply to the local agent and
get located, as he will have no recognized title, and his occupation of the
land will not count until this action has been taken.

Upon completion of his location, the locatee may enter upon and
occupy his land, and may commence his improvements ; and the Regulations require him to do so within one month.

The locatee will not be entitled to his patent until the expiration of
three years from the date of location, and he must then make proof that
the settlement duties have been fully completed. The settlement duties
required for each location are as follows, viz.:—

(*1*) *To have at least fifteen acres cleared and had under cultivation,*
of which two acres at least are to be cleared and cultivated annually
during the three years ;

(*2*) *To have built a habitable house, at least 16 by 20 feet in size ;*

(*3*) *And to have actually and continuously resided upon and culti-*
vated the land for three years after location.

A locatee is not bound to remain on the land all the time during the
three years ; but may be absent on business or at work for, in all, not
more than six months in any one year. He must, however, make it his
home, and clear and cultivate the quantity of land required (two acres at
least) each year.

Where a locatee holds two lots he may make the requisite improvements on either one or both, as he finds it most convenient.

A locatee who purchases an additional 80 acres under the Regulations
must, within three years from the date of sale, clear fifteen acres thereon,
and cultivate the same, before he will be entitled to the patent ; but he
is not required to build a house or reside on the purchased lot, where he
holds it in connection with a Free Grant.

The proof of the performance of the settlement duties must be: the
affidavit of the locatee himself, supported by the testimony of at least two
disinterested parties, which affidavits are to be filed with the local agent—
who, if satisfied as to the correctness of the statements contained therein,
recommends the issue of the patent, and transmits the application to the
Department. (See Form No. 4, in Appendix).

In case a locatee has, after the issue of his patent, absolutely and in good faith parted with the land patented to him as a Free Grant, he may take up another location by applying to the local agent, and making affidavit setting out the facts.

In case the locatee fails to perform the settlement duties required by law, his location is liable to forfeiture, and may be cancelled by the Commissioner of Crown Lands. Applications for cancellation must be made through the local agent, and be supported by the affidavits of the applicant and at least two credible witnesses, who will show what the present position of the lot is: whether the locatee ever occupied or improved, and, if so, to what extent, and the value of the improvements; when he ceased to occupy; and his address, if known. Upon receipt of this evidence the agent will, if he can ascertain the address of the locatee, notify him of the application, and call upon him to disprove the allega-tions, or show cause why his location should not be cancelled within thirty days. At the expiration of that time the agent will transmit the evidence, with anything he may have received from the locatee in reply, and his own report to the Department. (See Form No. 5.)

The assignment or mortgage of a homestead from a locatee to another party before the issue of his patent is invalid, and cannot be recognized by the Department. This does not, however, apply to the devise of a Free Grant lot by will, nor to transfers of land by a locatee for church, cemetery or school purposes, or the right of way of railroads.

All pine trees and minerals on land located or sold under the Free Grants Act are reserved from the location or sale, and are the property of the Crown; and the Commissioner of Crown Lands may at any time issue a license to cut the pine on such land. The locatee may, however, cut and use such pine trees as he requires for building and fencing on his land, and may also cut and dispose of any pine trees he meets within the actual process of clearing his land for cultivation; but any trees so dis-posed of are subject to the payment of the same dues as are payable by license-holders.

Holders of timber licenses have the right to haul their timber over the uncleared portion of any land located or sold, and to make such roads as may be necessary for the purpose, and to use all slides, portages and roads, and to have free access to all streams and lakes.

The Crown reserves the right to construct on any land located or sold, any Colonization Road, or deviation from the Government allowance for road; and to take from such land, without compensation, any timber, gravel or material required for the construction or improvement of any such road.

Any conveyance, mortgage or alienation (except a will) of the land located, by a locatee after the issue of patent and within twenty years from location, will be invalid unless it be by deed in which his wife is one of the grantors, and unless it be duly executed by her.

The land while owned by the locatee, his widow or heirs, shall be exempt from liability for debt during twenty years from the date of location. This exemption does not, however, extend to a sale for taxes legally imposed.

When a Free Grant locatee dies before the completion of his title, his representatives may continue the settlement duties and obtain a patent at the proper time upon filing the requisite proof. If he dies before the 1st July, 1886, intestate, evidence is required of the date of death and that he died intestate, giving the name of his widow, and the number and names of all his children, and if he left no wife or children, the name of his heir must be given; if he made a will, it must be sent in with proper proof of due execution according to law. If he died after the 1st July, 1886, probate or letters of administration to the real and personal estate, as the case may be, must be sent.

Where a locatee dies, whether before or after issue of patent, leaving a widow, she is entitled to the land during her widowhood in lieu of dower, unless she prefers to take her dower instead.

In making application for land, and in filing proof in support of applications for cancellation of a location, or for issue of patent, the applicant will save time and unnecessary trouble by filing his papers with, or mailing them to, the Crown Land Agent in whose agency the land is situated, as on account of the agent's local knowledge of the lands he has to deal with, the Department requires that his certificate be attached to all such applications.

Lands located or sold under the Rainy River Free Grants and Homesteads Act, or the regulations made thereunder, are liable to taxation from the date of such location or sale, and where taxes assessed on such land are in arrears for three years, the interest of the locatee or purchaser may be sold in the manner prescribed by law. When the tax-purchaser receives his deed, unless legal proceedings be taken to question it by some person interested within two years from the date of sale, he acquires the right and interest of the locatee or purchaser, and may obtain a patent on completion of the original conditions of location or sale.

In order to have his claim recognized, a tax-purchaser should file his deed in the Department, and two years after the date of the sale for taxes, should file evidence showing that no action has been taken to question his title, that there is no adverse claim on the ground of occupation or improvements, and that all arrears of taxes have been paid since he purchased. (See Revised Statutes (1887), Cap. 193, sections 159, 160 and 171, and Cap. 24, section 18). And in order to obtain a patent for the land, as a free grant, the tax-purchaser must also show that he has performed the settlement duties required by the Rainy River Free Grants and Homesteads Act, and that he has not already received the benefit of the said Act, or if he has received a grant of all the land which it allows him, that he has *bona fide* and absolutely parted with the same.

How to Purchase Public Lands.

In case a party should desire to purchase public land which has been surveyed, but is not within the jurisdiction of any Crown Land agent, he should make his application direct to the Department, and support it by the affidavits of at least two credible and disinterested persons. These affidavits should set out all facts in connection with the land which he seeks to purchase, and especially whether it has ever been occupied, whether occupied at the time the application is made, and, if so, by whom, and when such occupation commenced; whether any improvements have been made on said land, and, if so, the nature and extent of the same, and by whom and when they were made; and also, whether there is any claim made thereto adverse to that of the applicant, and based on the ground of occupation or improvements. If the applicant has acquired the interest or claim of some previous occupant, he should show the fact and file an assignment.

PART IV.

FORMS OF AFFIDAVITS

REFERRED TO IN THE FOREGOING ACTS.

0

SCHEDULE.

Of certain Acts and parts of Acts passed by the Ontario Legislature which are applied to the District of Rainy River.

REFERENCE TO ACT.	TITLE OF ACT.	EXTENT OF APPLICATION.
R. S. O., c. 7	An Act respecting the Territorial Districts of Muskoka, Parry Sound and Thunder Bay....	Sections 4 to 8, 11, 12, 14, 18, 19, 20 and 22 to 26.
R. S. O., c. 46,....	The Surrogate Courts Act......................	Secs. 10, 11, 12 and 13. Subject to the provisions of sec. 4, of this Act.
R. S. O., c. 175....	An Act respecting the establishment of Municipal Institutions in the Districts of Algoma, Muskoka, Parry Sound, Nipissing and Thunder Bay.	The whole, so far as now in force.
43 Vic., c. 12	An Act respecting the Administration of Justice, in the Districts of Algoma, Thunder Bay and Nipissing	Sub-secs. 5, 6, 7 and 8 of sec. 12.
45 Vic., c. 7.......	An Act relating to Division Courts in the Districts of Nipissing, Muskoka, Parry Sound and Thunder Bay, and to amend the Division Courts Acts.	The whole.
46 Vic., c. 23......	An Act respecting appeals to Stipendiary Magistrates from Municipal Assessment in Algoma, Muskoka, Parry Sound, Nipissing and Thunder Bay.	The whole.
47 Vic., c. 33	An Act to amend the Revised Statutes respecting Municipal Institutions in Algoma, Muskoka, Parry Sound, Nipissing and Thunder Bay.	The whole.

FORMS OF AFFIDAVITS USED IN APPLICATIONS FOR FREE GRANT AND OTHER CROWN LANDS.

No. 1.—AFFIDAVIT FROM A SINGLE MAN FOR 100 ACRES.

Set out the name, last place of residence, and occupation in full.

I,   of the   in the   make oath and say :

1. That I have not heretofore been located for any land under the " Free Grants and Homesteads Act," (except   ); nor have I obtained a Patent for any land as a Free Grant or any benefit under that Section of the said Act which provides for the remission of arrears due to the Crown by settlers who purchased in Free Grant Townships *(except for lot number but that I have absolutely and in good faith parted with the said land so patented to me, and I am entitled to and desire to obtain another location.)*

2. That I am of the age of   years.

3. That I desire to be located for lot number   in the   concession of the township of

4. That I believe the said land is suited for settlement and cultivation and is not valuable chiefly for its mines, minerals or pine timber; and that such location is desired for my benefit, and for the purpose of actual settlement and cultivation of such land, and not either directly or indirectly for the use or benefit of any other person or persons whatsoever, nor for the purpose of obtaining, possessing or disposing of any of the pine trees, growing or being on the said land, or any benefit or advantage therefrom, or any gold, silver, copper, lead, iron, or other mines or minerals, or any quarry or bed of stone, marble or gypsum thereon.

5. And that the said lot is wholly unoccupied and unimproved (*except*

Sworn before me, at }
this   day of   18. }

No. 2.—AFFIDAVITS WHERE APPLICANT IS THE MALE OR SOLE FEMALE, HEAD OF A FAMILY.

Set out the name, last place of residence, and occupation in full.   I,   of the   in the   make oath and say :

1. That I have not heretofore been located for any land under the "Free Grants and Homesteads Act," (*except* .   );
nor have I obtained a Patent for any land as a Free Grant or any benefit under that Section of the said Act which provides for the remission of arrears due to the Crown by settlers who purchased in Free Grant Townships (*except for lot number but that I have absolutely and in good faith parted with the said land so patented to me, and I am entitled to and desire to obtain another location.*)

2. That I am the male (or) *sole female* head of a family, having   children under eighteen years of age, residing with me, consisting of   son and   daughter .

3. That I desire to be located under the said Act, and the Regulations made thereunder for lot number   in the   concession, and lot number in the   concession of the township of

4. That I believe the said lands are suited for settlement and cultivation, and are not valuable chiefly for their mines, minerals, or pine timber.

5. That such location is desired for my benefit, and for the purpose of actual settlement and cultivation of such lands, and not either directly or indirectly for the use or benefit of any other person or persons whatsoever, nor for the purpose of obtaining, possessing or disposing of any of the pine trees growing or being on the said lands, or any benefit or advantage therefrom, or any gold, silver, copper, lead, iron, or other mines or minerals, or any quarry or bed of stone, marble or gypsum thereon.

6. And that the said lots are wholly unoccupied and unimproved (*except*

Sworn before me, at }
this   day of   18 . }

We,   of the   in the   and
of the   in the   each for himself, make oath and say : that
I am well acquainted with   named in the above affidavit, and that he
is the   head of a family and has   children, under eighteen years
of age, (consisting of   son and   daughter ,) residing with him ; and I further
make oath and say that I know lots number   , in the   concession of
the township of   referred to above, that I am not aware of any claim to the
said lots on the grounds of occupation, improvements or otherwise, adverse to that of
the applicant, and that the said lots are wholly unoccupied and unimproved (*except*

Sworn before me, at .   }
this   day of   18 . }

No. 4.—APPLICATION FOR PATENT UNDER THE PROVISIONS OF "THE FREE GRANTS
AND HOMESTEADS ACT."

To the Crown Lands Agent:

SIR,—I have the honour to apply, under the provisions of "The Free Grants and
Homesteads Act," for a Patent from the Crown for my Homestead, upon the grounds
set forth in the following affidavits, and have to request that the said Patent, when
issued, be mailed to the following address, viz. :

Dated this   18   .

Affidavit of Applicant.

Ontario, ⎱ I,   of the
of   ⎰ in the   of
To Wit :   ⎰ make oath and say :—

Recommendation for Patent this 18 — Crown Land Agent.

1. That I desire to obtain my Patent under the provisions of the Eighth
Section of "The Free Grants and Homesteads Act," for lot
of the township of   for which lot I was located on the
day of   18   .

2. That since then I have been an actual resident upon, and have cultivated the said lot continuously for   years, and that I am still residing
upon and cultivating the same.

3. That I have cleared upon the said lot, and had under cultivation last
season   acres at least, and that I have erected buildings thereon of
the following descriptions and dimensions, viz. : A house fit for habitation
x   feet at least,

4. That I have not been located for any other land (except   )
nor have I obtained Patent for any land, as a Free Grant, or by remission of
arrears, under the provisions of the said Act ; and that I am well entitled to
the Patent for the said lot, and am not aware of any adverse claim thereto on
the grounds of occupation, improvements or otherwise.

Sworn before me, at   ⎱
this   day of   18   . ⎰

Affidavit in Support of Application

Ontario, ⎱ We,   of the township of   in the
of   ⎰ of   and   of the same place, yeoman, each for
To Wit :   ⎰ himself, make oath and say :—

That I know lot   in the   concession of the township of
described in the affidavit of   the Applicant for Patent ; that the said affidavit
has been read over to me, and that all the statements made therein respecting the
residence of the said   on the said lot, and the cultivation and improvements made by him thereon are true in substance and in fact, and that I am not
aware of any adverse claim thereto.

Sworn before me, at   ⎱
this   day of   18   . ⎰

No. 5.—APPLICATION FOR CANCELLATION OF A LOCATION.

Affidavit of Applicant.

Ontario,   ⎱ I,   of the township of   in the district
District of ⎰ of   , yeoman, make oath and say :—
To Wit :   ⎰ 1. That I desire to be located for lot number   in the
concession of the township of

2. That I am informed that the said lot   located on the   day of
A.D. 18   , to one

3. That I know the said lot, and personally visited and examined on the ; and that there was no person at that time residing thereon ; and that I did not discover any improvements whatever on the said lot ; and that from said examination, and from information which I have received, I verily believe that the said locatee has never occupied or improved the said lot ;

If locatee has occupied or improved at any time set out when he ceased to occupy, what improvements he made, when they were made, and in what position the lots are at time of application.

4. That so far as I am aware the said locatee is not occupying or improving any other land in the said township, and resides at present at

5. And that I have not, neither has any person for me, either directly or indirectly by purchase or otherwise from the said locatee, or any other person, acquired any interest in the said lot.

Sworn before me, at
in the of this
day of A D. 18 .

Affidavit in Support of Application.

Ontario, } We, · of the township of in the district
District of } of and f the same place, yeomen, each for
To Wit : } himself, make oath and say :—

1. That I know lot number in the concession of the township of which located to and that I personally visited and examined the said lot on the , that there was no person then residing thereon, nor were there any improvements whatever ; and that from said examination and from information which I have received, I verily believe that the said locatee has never occupied or improved the said lot ;

If locatee has occupied or improved at any time set out when he ceased to occupy, what improvements he made, when they were made, and in what position the lots are at time of application.

2. That the said locatee is not, as far as I am aware, occupying or improving any other land in the said township, and that he resides at

Sworn before me, at
in the of this
day of A.D. 18 .

I hereby certify that I have no reason to doubt the statements contained in the foregoing affidavits ; and also that I did on the day of mail to the locatee of said lots at Post Office, a letter notifying him of the application for cancellation, and calling upon him to show cause why it should not be allowed, and since then I have not received any reply to the said notice, except

Crown Land Agent.

47

No. 6.—AFFIDAVIT IN SUPPORT OF APPLICATION FOR LAND UNDER THE MINING ACT.

Ontario,   }   I,   of the township of   in the District
District of   } of   and I,   of the township of
To Wit :   } in the district of   do solemnly swear :—

1. That on the   day of   I personally visited and carefully examined   lot number   in the   concession of the township of   and at that time there was no person residing on said lot, and there were no improvements thereon.

2. That there was no visible trace nor indication of work having been done on said lot by any person or persons for mining or other purposes.

3. And that to the best of my knowledge and belief there is no claim to said lot by any person or persons adverse to that of   the applicant on the ground of priority of discovery of mineral thereon, or otherwise.

Sworn before me, at
in the district of   this
day of   A.D. 18 .

No. 7.—AFFIDAVIT TO BE TAKEN BY A MALE OR FEMALE HEAD OF A FAMILY WHO DESIRES TO PURCHASE LAND, SUBJECT TO SETTLEMENT IN NORTH NIPISSING AND ALGOMA DISTRICTS.

Canada,   }   I,   of   being the
Province of Ontario, } head of a family, and desirous of purchasing lot number
District of   } in the   concession of the township of   make
To Wit :   } oath and say :—

1. That I am the   head of a family.

2. That the said land is wholly unoccupied and unimproved, and I believe the same to be suited for settlement and cultivation.

3. That I desire to purchase the said land for the purpose of settling thereon and for cultivation, and not for speculative purposes or for the cutting or disposing of any timber there may be thereon.

Sworn before me, at
this   day of   18 .

No. 8.—AFFIDAVIT TO BE TAKEN BY A MALE PERSON ABOVE THE AGE OF EIGHTEEN YEARS, AND NOT THE HEAD OF A FAMILY, WHO DESIRES TO PURCHASE LAND AND BECOME A SETTLER IN SAME DISTRICTS.

Canada,   }   I,   of   make
Province of Ontario, } oath and say :—
District of   } 1 That I am of the full age of eighteen years.
To Wit: } 2 That I am desirous of purchasing lot number   in
the   concession of the township of   in the district of
and that the said land is wholly unoccupied and unimproved, and I believe the same is suited for settlement and cultivation.

3. That I desire to purchase the said land for the purpose of settling thereon and for cultivation, and not for speculative purposes or for the cutting or disposing of any timber there may be thereon.

Sworn before me at
this   day of   18

No. 9.—Affidavit in Support of Application for Patent for Lands Sold Subject to Settlement.

Ontario,   ⎫ We, of the township of in the
District of   ⎬ of and of the same place, yeomen, each for
To Wit: ⎭ himself, make oath and say :—

1. That I know lot number in the concession of the township of

2. That there are acres cleared and had under cultivation and crop on the said lot.

3. That the following buildings have been erected thereon, namely :—A house fit for habitation x feet.

4. That said lot is occupied by and has been continuously occupied by for years, and the improvements made thereon were made by

5. That I do not know of any claim to or occupation of said lot adverse to that of

Sworn before me, at
this day of 18 . ⎱

Recommended for Patent this 18 . Crown Land Agent.

PART V.

MINERAL PROSPECTS.

EXTRACTS FROM THE GENERAL MINING ACT.

4 (R.R.)

is
ni
go
bo
ric
(1£
W
the
roc
ful

MINERAL PROSPECTS.

The mineral wealth of this district, although less easily estimated, bids fair to be even greater than that of the timber wealth. The results thus far of prospecting establish the existence of gold, silver, copper and iron, and it is more than probable that the valuable veins discovered in Algoma will be found to extend through the Rainy River district.

That the geological formation is indicative of valuable mineral deposits is verified by Professor Bell's geological surveys. A band of rocks running south-west from Lake Shebandowan (in neighbourhood of which gold has been found in considerable quantities,) to the international boundary, and thence to Lake Vermillion in Minnesota, is also said to be rich in auriferous deposits. The Public Works Report of the Dominion (1875), says that, "The Indians, both of Rainy Lake and Lake-of-the-Woods have among them specimens of native gold and silver ore, which, they affirm, is to be found in places known to them in abundance, and the rock formation is such as to corroborate their statement. Iron ore is plentiful in many sections, and charcoal for smelting easily obtainable. Granite, which report says is equal in texture and fitness to the best imported specimens, is to be found at the Lake-of-the-Woods, and the steatite, of which the Indians make pipes, a very valuable article for the construction of furnaces, is quite abundant at Rainy Lake and Sabaskin. It was stated in evidence before the Committee on Immigration and Colonization at Ottawa, in 1878, that coal had been discovered in the vicinity of Rainy River. There does not appear to be any reason, on scientific grounds, for doubting the existence of coal in that region, but its quality or the extent of the deposits, if they exist, are subjects for further inquiry before much reliance can be placed on the value of the alleged discovery." The information thus afforded, while not absolutely conclusive, is so far indicative of mineral deposits of greater or less richness in the region we have been describing, as to suggest the propriety of a careful exploration, with the special object of ascertaining more thoroughly the value of the district for mining purposes.

The reduction works being built at Rat Portage have a capital stock of $200,000, and will treat thirty tons of ore daily. The works will cost $60,000, and Rat Portage gave a bonus of $10,000. These works are to serve two purposes, namely, to treat not only all kinds of gold ores, but silver ores as well. The mill is capable of treating thirty tons of gold ores per day, and will be able to treat an equal quantity of silver ores without hindering that part of the mill running on gold.

Mineral Lands.

Public lands, which are open for sale, may be sold under " The General Mining Act " (Revised Statutes of Ontario, Cap. 31), at the rate of two dollars per acre, cash, in the Rainy River District. The patent is issued upon payment, and contains a reservation of all pine trees standing or being on the land. The pine continues to be the property of the Crown, who may at any time issue a license to cut it, and the party holding the license is empowered to enter at all times upon the land, cut and remove it, and make all necessary roads for that purpose. (See extracts from Mining Act which follow.)

Applications to purchase land under the Mining Act should be made to the Crown lands agent in the district, and should be accompanied by the purchase money, together with affidavits of at least two credible and disinterested parties, showing that the land is unoccupied and unimproved (except by or on behalf of the applicant), and that there is no claim thereto adverse to his on the ground of occupation, improvements or otherwise. (See Form in Appendix, No. 6).

The affidavits necessary for use in the purchase of mineral lands will also be found in Part IV.

53

EXTRACTS FROM THE GENERAL MINING ACT.

REVISED STATUTES ONTARIO, 1887, CHAPTER 31, WITH AMEND-
MENTS OF 1890 INCORPORATED THEREIN.

*The following are the principal clauses of the General Mining
Act and amendments thereto.*

5. No reservation or exception of gold, silver, iron, copper or other mines or minerals, shall be inserted in any patent from the Crown granting lands in this Province sold as mining lands. R. S. O. 1877, c. 29, s. 5. *No reserva- tions in pat- ents of mining lands.*

6. Any person or persons may explore for mines or minerals on any Crown Lands, surveyed or unsurveyed, and not for the time being marked or staked out and occupied as hereinafter mentioned. R. S. O. 1877, c. 29, s. 6. *Crown lands may be ex- plored for mines, etc.*

8. Such lands so sold, when situate in unsurveyed territory, or in Townships surveyed in sections, shall be sold in blocks to be called "mining locations." R. S. O. 1867, c. 29, s. 8. *Mining loca- tions.*

9. Mining locations under this Act shall conform to the fol- lowing requirements :— *Form and size of mining loca- tions.*

1. "In the unsurveyed territory within the districts of Algoma, Thunder Bay and Rainy River and that part of the District of Nipissing which lies north of the French River, Lake Nipissing and the River Mattawa," every regular mining location shall be rectangular in shape, and the bearings of the out- lines thereof shall be due north and south and due east and west astronomically ; and such location shall be of one of the following dimensions, namely, eighty chains in length by forty chains in width, containing three hundred and twenty acres, or forty chains square, containing one hundred and sixty acres, or forty chains in length by twenty chains in width, containing eighty acres or twenty chains in length by twenty chains in width containing forty acres. *Territory bor- dering on lakes Superior and Huron, French River, etc.*

2. Where a mining location in the unsurveyed lands in the territory aforesaid, borders upon a lake or river a road allow- ance of one chain in width shall be reserved along the margin of the lake or river ; and the width of the location shall front on the road allowance ; and the bearings of the other out- lines of the location shall be due north and south, and due east and west astronomically ; and the location shall otherwise conform to the requirements of the preceding sub-section as nearly as the nature of the land will admit. *When loca- tions border on lakes and rivers in said territory.*

When in townships in said territory surveyed in sections.
3. In the townships in the said territory surveyed, or hereafter to be surveyed in sections, every mining location, after such survey, shall consist of a half section, a quarter section, or an eighth of a section or a sixteenth of a section.

Reservation for roads.
4. In all patents for mining locations in the territory aforesaid, there shall be a reservation for roads of five per centum of the quantity of land professed to be granted.

Locations in other unsurveyed territory.
5. In the unsurveyed lands not situate within the limits of the territory aforesaid, mining locations shall be as may be defined by any Order in Council hereafter to be made. R. S. O. 1877, c. 29, s. 9.

How mining locations in unsurveyed territory to be surveyed.
10. Mining locations in unsurveyed territory shall be surveyed by a Provincial Land Surveyor, and be connected with some known point in previous surveys, or with some other known point or boundary (so that the tract may be laid down on the office maps of the territory in the Crown Lands Department), at the cost of the applicants, who shall be required to furnish, with their application, the surveyor's plan, field notes and description thereof, shewing a survey in accordance with this Act, and to the satisfaction of the Commissioner of Crown Lands. R. S. O. 1877, c. 29, s. 10.

Price of locations.
11. The price of all Crown Lands to be sold as mining locations in the said territory, mentioned in sub-section 1 of section 9 of this Act, shall be $2 per acre. R. S. O. 1877, c. 29, s. 11; 49 V. c. 8, s. 1.

Pine trees reserved.
12. The patents for all Crown Lands, hereafter to be sold as mining lands, shall contain a reservation of all pine trees standing or being on the said lands, which pine trees shall continue to be the property of Her Majesty; and any person now or hereafter holding a license to cut timber or saw logs on such lands, may, at all times, during the continuance of the license, enter upon such lands, and cut and remove such trees, and make all neces-

Patentees may use timber for building, fencing, etc., on the land.
sary roads for that purpose; but the patentees, or those claiming under them, may cut and use such trees as may be necessary for the purpose of building, fencing and fuel, on the land so patented, or for any other purpose essential to the working of the mines thereon; and may also cut and dispose of all trees required to be removed in actually clearing the land for cultivation; but no pine trees (except for the said necessary building, fencing and fuel, or other purpose essential to the working of the mine), shall be cut beyond the limit of such actual clearing; and all pine trees so cut and disposed of (except for the said necessary building, fencing and fuel, or other

Timber cut to be subject to dues.
purpose aforesaid), shall be subject to the payment of the same dues, as are at the time payable by the holders of licenses to cut timber or saw logs. R. S. O. 1877, c. 29, s. 12.

* * * * * * *

42.—(1) The Lieutenant-Governor may, from time to time, Appointment appoint local officers or agents to receive applications for the of agents to sale of mining lands in their respective agencies and to carry sell mining lands. out the provisions of any regulations and Orders in Council in that behalf and to supply information to intending purchasers and they shall be paid in such manner and at such rates as the Lieutenant-Governor in Council may direct.

(2) The Lieutenant-Governor in Council may, from time to time, make such regulations as he deems necessary or expedient for the purpose of carrying out this section.

PART VI.

THE RAINY RIVER FIRE DISTRICT.

AN ACT TO PRESERVE THE FORESTS FROM DESTRUCTION BY FIRE.

5 (R.R.)

THE RAINY RIVER FIRE DISTRICT.

The Rainy River District is included in the Fire District No. 2 (See Cap. 22, R. S. O.,) the latter being described as follows :—

District No. 2.—All that part of the said Province lying west of Provincial Land Surveyor Albert P. Salter's meridian line between ranges twenty-one and twenty-two west, near Bruce Mines, in the District of Algoma, and west of the said meridian line produced, to the northern boundary of the Province, the said meridian line being the western boundary of the Fire District established by the Proclamation of March 27th, 1878.

AN ACT TO PRESERVE THE FORESTS FROM DESTRUCTION BY FIRE.

CHAPTER 213, REVISED STATUTES OF ONTARIO, 1887.

PROCLAMATION OF FIRE DISTRICT, ss. 1-3.
RESTRICTIONS AS TO STARTING FIRES, s. 4.
PRECAUTIONS AS TO FIRES FOR CLEARING LAND, s. 5.
Fire for cooking, s. 6.
Matches, cigars, firearms, s. 7.
ACT TO BE READ TO EMPLOYEES BY SURVEYORS, ETC., s. 8

LOCOMOTIVE ENGINES, MANAGEMENT OF, ss. 9, 10.
PENALTY, s. 11.
LIMITATION OF ACTIONS, s. 12.
DISPOSAL OF FINES, s. 13.
ENFORCING ACT, s. 14.
RIGHT TO DAMAGES NOT AFFECTED, s. 15.

HER MAJESTY, by and with the advice and consent of the Legislative Assembly of the Province of Ontario, enacts as follows :—

1. The Lieutenant-Governor, may, by proclamation to be made by him from time to time, issued by and with the advice and consent of the Executive Council, declare any portion or part of the Province of Ontario, to be a fire district. 41 V. c. 23, s. 1. *Lt.-Governor may proclaim a fire district*

2. Every proclamation under this Act shall be published in the *Ontario Gazette*; and such portion or part of the Province as is mentioned and declared to be a fire district in and by the said proclamation, shall, from and after the said publication, become a fire district within the meaning and for the purposes of this Act. 41 V. c. 23, s. 2. *Publication of fire district.*

3. Every portion or part of the Province mentioned in the proclamation shall cease to be a fire district upon the revocation by the Lieutenant-Governor in Council of the proclamation by which it was created. 41 V. c. 23, s. 3. *Revocation.*

Fires not to be started except for certain purposes and in certain periods.

4. It shall not be lawful for any person to set out, or cause to be set out or started, any fire in or near the woods within any fire district between the 1st day of April and the 1st day of November in any year, except for the purpose of clearing land, cooking, obtaining warmth, or for some industrial purpose; and in cases of starting fires for any of the above purposes, the obligations and precautions imposed by the following sections shall be observed. 41 V. c. 23, s. 4.

Precautions to be taken in case of clearing land.

5. Every person who shall, between the 1st day of April and the first day of November, make or start a fire within a fire district for the purpose of clearing land, shall exercise and observe every reasonable care and precaution in the making and starting of such fire, and in the managing of and caring for the same after it has been made and started, in order to prevent the fire from spreading and burning up the timber and forests surrounding the place where it has been so made and started. 41 V. c. 23, s. 5.

Precautions in case of cooking, etc.

6. Every person who shall, between the 1st day of April and the 1st day of November, make or start within such a district a fire in the forest, or at a distance of less than half-a-mile therefrom, or upon any island, for cooking, obtaining warmth, or for any industrial purpose, shall—

1. Select a locality in the neighbourhood in which there is the smallest quantity of vegetable matter, dead wood, branches, brushwood, dry leaves, or resinous trees;

2. Clear the place in which he is about to light the fire by removing all vegetable matter, dead trees, branches, brushwood, and dry leaves from the soil within a radius of ten feet from the fire;

3. Exercise and observe every reasonable care and precaution to prevent such fire from spreading, and carefully extinguish the same before quitting the place. 41 V. c. 23, s. 6.

Precautions in cases of matches, burning substances, etc.

7. Any person who shall throw or drop any burning match, ashes of a pipe, lighted cigar or any other burning substance, or who shall discharge any firearm within such fire district, shall be subject to the pains and penalties imposed by this Act if he neglect completely to extinguish before leaving the spot the fire of such match, ashes of a pipe, cigar, wadding of the firearm or other burning substance. 41 V. c. 23, s. 7.

Act to be read to employees by heads of surveys, lumberers, etc.

8. Every person in charge of any drive of timber, survey or exploring party, or of any other party requiring camp-fires for cooking or other purposes, within a fire district, shall provide himself with a copy of this Act, and shall call his men together and cause the Act to be read in their hearing, and explained to them at least once in each week during the continuance of such work or service. 41 V. c. 23, s. 8.

Precautions as to locomotives.

9. All locomotive engines used on any railway which passes through any fire district or any part of a fire district, shall, by the company using the same, be provided with and have in use all

the most approved and efficient means used to prevent the escape of fire from the furnace or ash-pan of such engines, and the smoke-stack of each locomotive engine so used shall be provided with a bonnet or screen of iron or steel wire netting, the size of the wire used in making the netting to be not less than number nineteen of the Birmingham wire gauge, or three sixty-fourth parts of an inch in diameter, and shall contain in each inch square at least eleven wires each way at right angles to each other, that is in all twenty-two wires to the inch square. 41 V. c. 23, s. 9.

10. It shall be the duty of every engine driver in charge of Duty of engine drivers. a locomotive engine passing over a railway within the limits of any fire district, to see that all such appliances as are above mentioned are properly used and applied, so as to prevent the unnecessary escape of the fire from any such engine as far as it is reasonably possible to do so. 41 V. c. 23, s. 10.

11. Whosoever unlawfully neglects or refuses to comply Penalty. with the requirements of this Act in any manner whatsoever, shall be liable, upon a conviction before any Justice of the Peace, to a penalty not exceeding $50 over and above the costs of prosecution, and in default of payment of such fine and costs the offender shall be imprisoned in the common gaol for a period not exceeding three months; and any railway company permitting a locomotive engine to be run in violation of the provisions of section 9 of this Act shall be liable to a penalty of $100 for each offence, to be recovered with costs in any Court of competent jurisdiction. 41 V. c. 23, s. 11.

12. Every action for any contravention of this Act shall be Limitation of actions. commenced within three months immediately following such contravention. 41 V. c. 23, s. 12.

13. All fines and penalties imposed and collected under this Disposal of fines. Act shall be paid one-half to the prosecutor and the other half to Her Majesty for the public use of the Province. 41 V. c. 23, s. 13.

14. It shall be the special duty of every Crown Land Government agents to enforce this Act. agent, woods and forest agent, free grant agent and bush ranger to enforce the provisions and requirements of this Act, and in all cases coming within the knowledge of any such agent or bush ranger to prosecute every person guilty of a breach of any of the provisions and requirements of the same. 41 V. c. 23, s. 14.

15. Nothing in this Act contained shall be held to limit or Act not to interfere with right of action for damages. interfere with the right of any party to bring and maintain a civil action for damages occasioned by fire, and such right shall remain and exist as though this Act had not been passed. 41 V. c. 23, s. 15.

6 (H.R.)